Management Division 2021

Held at the 2021 AIChE Spring Meeting and 17th Global Congress on Process Safety

Online
18 - 23 April 2021

ISBN: 978-1-7138-5338-1

Printed from e-media with permission by:

Curran Associates, Inc.
57 Morehouse Lane
Red Hook, NY 12571

Some format issues inherent in the e-media version may also appear in this print version.

Printed with permission by Curran Associates, Inc. (2022)

AIChE
120 Wall Street, FL 23
New York, NY 10005-4020

Phone: (800) 242-4363
Fax: (203) 775-5177

www.aiche.org

Additional copies of this publication are available from:

Curran Associates, Inc.
57 Morehouse Lane
Red Hook, NY 12571 USA
Phone: 845-758-0400
Fax: 845-758-2633
Email: curran@proceedings.com
Web: www.proceedings.com

TABLE OF CONTENTS

616879 Management Systems and Culture Important for Sustained Operations

Abstract:

Currently, the Department of Energy (DOE) stores approximately 56 million gallons of radioactive and hazardous waste in 177 underground tanks at the Hanford Site. The waste was generated from production of plutonium for the United States' nuclear weapons program starting in the second World War and finishing in 1990. Approximately 20 million gallons of that waste is in a liquid form (supernate), approximately 10 million gallons is in the form of insoluble sludge materials, and the remainder is in a partially soluble solid form referred to as saltcake. Treatment of the tank waste is planned with the Waste Treatment and Immobilization Plant (WTP) being the principle plant where this will be accomplished. The WTP is currently being constructed and large portions of the plant are already starting up. Waste treatment commissioning and operations are planned to start in the early 2020s.

The start of waste treatment operations will mark a significant event in Hanford history. Not since the 1980s, or at least a generation of engineers, will the Site experience the sustained operations of a multi-functional chemical process plant. This situation makes for an interesting study in transitioning management systems and culture from one supportive of design and construction to one conducive to sustained operations. Making ready for safe operations and then to demonstrably show readiness in Operational Readiness reviews and assessments is a well-defined process in the DOE Environmental Management complex. However, there is little programmatic guidance on how to make ready and demonstrably show readiness for sustained operations.

The development of a Sustained Operations Readiness Program was conducted based on the author's professional experience, input from specific individuals with relevant expertise and the author's review of public information. A Sustained Operations Readiness Program is not intended to replace or in any way diminish the role for Operational Readiness Reviews and Assessments as required by the Department of Energy. Rather, the intent is to build upon the Operational Readiness concept to maximize readiness for sustained operations through commissioning and into operations, and to establish a culture of simultaneous operations excellence in production, safety and environmental management. The work has concluded that the state of sustained operations for a complex processing plant can be objectively characterized by 12 features. These features represent the conducive culture or conditions in which the plant, paper and people interact to produce a sustained operations state. The 12 features were further consolidated into the five programmatic and executable elements described below with associated recommendations specific for Hanford waste treatment operations:

Risk management
Risks and opportunities should prioritize continuous improvement initiatives and be informed by the design and technical authorities and stakeholder monitors.
For Hanford waste treatment operations, an integrated risk management program led by the DOE is recommended.

Continuous improvement
All levels of the project organization should exhibit commitment to continuous improvement and learning.
At Hanford, the tank farm contractor had introduced the Lean management system as a continuous

improvement program. The program has yielded efficiencies in tank farm operations and waste treatment operations planning and introduced a culture of continuous improvement. A waste treatment operations program-wide continuous improvement system is recommended for Hanford waste treatment operations.

Continuity of design and technical authorities
The authorities are responsible for managing knowledge for its effective use in the other elements. Recommended for Hanford waste treatment operations is a body of technical program managers from the key technology providers chartered to provide the technical authority and a continuing role for the WTP Engineering, Procurement and Construction contractor as the design authority.

Stakeholder monitoring
Active and deliberate monitoring of stakeholder relationships.
Hanford waste treatment operations is characterized by complex and wide-ranging stakeholders at Federal, State and local levels of government and regulation and various public bodies. Centralized and integrated stakeholder monitoring and management is recommended as an organizational element.

Sustained operations integrated business planning
Business planning integrates and balances technical, financial and regulatory objectives. In a commercial venture this is arguably a business plan.
An integrated Hanford waste treatment operations business plan is recommended for the Site to most effectively compete with other sites in the current Federal environment competitive for Congressional funding.

Readiness evaluation for sustained operations is a continuous function that progresses from overseeing establishment of the five elements during design and construction phases to monitoring the output of those elements performing during commissioning and operations. Continuously assessing readiness for sustained operations becomes an important business function for the Department of Energy's Office of River Protection as competition for Federal funds increases with increasing waste treatment operations throughout the Environmental Management complex.

621580 Carbon Leadership in a Low Carbon World

Abstract:

Carbon dioxide levels in the atmosphere have increased from about 275 ppm prior to the industrial revolution to over 415 ppm today. Nature cannot keep up, and there is now over 2 trillion tonnes of CO_2 that has accumulated in the atmosphere. This presentation will touch on point source emissions capture but will focus on Direct Air Capture technology coupled with subsurface sequestration (DAC+S). A technology overview and a summary of potential business models will be discussed.

620242 Leading by Example – Innovations in Pharmaceutical Development to Position for a Sustainable Future

Abstract:

As the active substances and modalities for advanced therapies have become more structurally complex and R&D timelines continue to be compressed, the chemistry and technologies necessary to provide safe, robust and economical access to these molecules needs to keep pace. Enabling technologies such as continuous processing, automation and predictive modeling through first-principles, as well as machine learning can expedite process development, scale-up and technology transfer while ensuring enhanced process efficiency and delivering quality medicines to patients more effectively. The importance of strong leadership and novel concepts in pharmaceutical industry has become even more evident in recent Covid-19 pandemic crisis, different organizations are actively thinking of implementing these paradigms to their benefit. This presentation will share some of the leadership experiences and novel technologies that can be leveraged to accelerate process development and manufacturing and discuss the challenges and opportunities to position for a sustainable future.

620403 Enabling Innovation in New Modalities Synthesis

Abstract:

Dr. Cruz will speak on a subject related to leadership in the pharmaceutical industry.

620244 Getting Ahead of the Curve – Emerging Trends and Success Factors in Biopharmaceutical Development

Abstract:

This talk will discuss leadership aspects involved in pharmaceutical development.

618433 Imposter Syndrome

Abstract:

Imposter syndrome (IS) is a mindset wherein accomplished individuals feel that they have succeeded only due to luck or because others have not realized that they are frauds and thus not deserving of their success. This syndrome is an emotional response rather than a logical one in that the individual dealing with IS knows all details of his/her specific difficulties and struggles and therefore feels that they are undeserving of their achievements. IS is widespread and is characteristic of individuals irrespective of their experience level, gender, ethnicity, stature, and chosen field. IS causes us to avoid applying for certain jobs, negotiating effectively, and enjoying our accomplishments. In this talk, possible reasons that IS is prevalent will be described. Approaches to overcome this syndrome and to minimize the effects of IS on our careers, goals, and subordinates will be discussed

MBBR Wastewater Treatment System for Nitrification of Combined Industrial and Municipal Wastewaters under Cold Weather Conditions with Wide Load Fluctuations

Siva Angappan[1], Eugene Townsley[2] and Somnath Basu[1],

1. Headworks International, 11000 Brittmoore Drive, Houston, TX 77041
2. Batesville Wastewater Treatment Plant, Batesville, AR 72501

Abstract

Uncontrolled discharge of ammonia in treated effluent from wastewater utilities is harmful to the receiving bodies of water. At high concentrations ammonia is toxic to aquatic life. It also adversely impacts the presence of dissolved oxygen which is critical to the health of water bodies. Additionally, discharge of nitrogen in treated wastewater in all forms, including ammonia, promotes undesirable growth of aquatic plants e.g., hyacinth, algae, etc., also called eutrophication, which renders water unsuitable for use and recreational purposes. Environmental regulators throughout North America and globally are increasingly calling for effluent nitrogen control, which requires nitrification as the first step. Nitrification involves biological oxidation of ammonium ions in wastewater. It is kinetically a very slow process and sensitive to environmental conditions, especially wastewater temperature, pH, presence of heavy metals and other toxicants. Many treatment plants, especially in rural US and Canada are lagoon based that can remove only organic pollutants - BOD (biological oxygen demand). In winter, these lagoons are exposed to very low ambient temperatures. Nitrification is difficult to sustain under cold temperatures <5°C. Moving Bed Biofilm Reactors (MBBR) have been shown to effectively conduct nitrification under sustained low temperature conditions while consistently achieving low ammonia limits. Other benefit of the MBBR process include – (a) automatic growth and retraction of the biofilm under varying flows and loads without having the operator to adjust, monitor, or control parameters such as recycled activated sludge (RAS) and mixed-liquor suspended solids (MLSS), (b) requirement of much smaller reactor volume and footprint area, and (c) ability to quickly recover from hydraulic surges, and toxic and shock loads.

Batesville, AR, has a population of around 10,000 but its wastewater treatment plant treats 60,000 population equivalent flows due to many industrial facilities in the area including poultry and dairy industries. Prior to 2015, lagoons were used to treat the wastewater to remove only BOD. Lagoons were also used to settle the solids. Due to substantial growth in industrial discharges to the facility, the plant was stretched to the limit to remove BOD, especially in winter, when the water temperature in the lagoons reach as low as 4° C.

In 2015, the plant's permit was revised to remove ammonia-nitrogen with a year-around limit. Due to the very low water temperatures in winter expansion of the lagoons was not considered a viable option. MBBR was selected as the technology to meet the new regulatory requirements for BOD and ammonia-nitrogen control due to its ability to perform under very low temperatures and handle varying and toxic loads from industrial contributors.

This presentation will consist of an analysis of many years of operating data demonstrating how the MBBR process helped the treatment plant achieve its goals of producing stable effluent quality consistently over the year even under low temperatures and varying loads. Operating performance will be compared with kinetic models based on the principle of MBBR technology.

620741 Technical Women's Leadership Forum – What are we waiting for?

Abstract:

The Women's Leadership Forum (WLF) is one of the largest and oldest employee resource networks (ERN) at 3M. Yet until only recently, relatively few women at the bench (technical track) participated in this ERN. In addition, women reaching upper level technical positions such as Corporate Scientist or Division Staff Scientist, remained disappointingly low over the past several decades. In 2018, female scientists responded by creating a technical chapter (T-WLF) in order to specifically address the needs of technical women. Core WLF pillars of "I believe in Me", "I believe in You", and "We believe in Us" were reinterpreted to meet the needs of technical women. The journey began with the creation of deliberate efforts to 1- *Retain* existing technical employees with Career Development and Technical Men as Advocates committees, 2- Implement best practices and build a better pipeline with *Outreach* committee, and 3- Address the early gap for girls in STEM with *Inspire* committee. In short two years, team has accomplished to collaborate with executive management and amplified the voice of the female technical community, partnered with 3M's Office of Intellectual Property to conduct surveys and focus groups, successfully granted funding from 3M to two non-profit organizations, highlighted successful technical women stories on internal internet, increased presence and engagement with recruiting events, and initiated global expansion. Today, 3M technical women are not waiting; they are part of making change happen.

In 2018, Mike Roman (3M Chairman and CEO) established the 4th pillar of the 3M Strategy, titled "People and Culture" with the strategic goal of strengthening our competitive edge in attracting top talent as well as developing our people & advancing our culture. The three prongs to this pillar include: inclusive workplace, development for all, and future of work (flexibility/global/digital). To continue to build on the strength of 3M's inclusive culture, the CEO Inclusion Council (headed by Mike) was established which includes the Employee Resource Networks' leaders along with Inclusion Leaders in division & staff groups. I was named the Inclusion Leader for the Corporate Research Division in late 2018 and will discuss a variety of Case Studies (best practices and key learnings) from my experience over the last two years.

In my talk, I will cover three different areas: psychological safety, personal advocacy, and organizational leadership. I will discuss my own journey from awareness to allyship to advocacy. I will present case studies including what you can do: mentorship, speaking out for yourself and others, moving beyond your company and into the community as well as driving metrics that facilitate change. Finally, my presentation is meant to help others create an environment where people can show up as their authentic selves and flourish in a diverse, inclusive and equitable environment.

621648 Case Study: EDI in Practice

Abstract:

Part 2 of the session is a case study from a plant manager of DuPont, Jana Green, who took action with her employees after the George Floyd murder, to foster small group discussions. Her advanced EDI practices illustrate how she created psychological safety for the employees to share their feelings, ways she facilitated the group discussions with an approach that was mindful of different stakeholder groups. She exhibited high emotional intelligence throughout the discussions and went on to share them operations leaders across the U.S.

621135 *Great Engineering Skills and High Emotional Intelligence: A Winning Combination*

Abstract:

Every one of us in the field of chemical engineering has invested many hours in the learning and application of principles of *chemistry*, biology, physics, and math to solve problems that involve the production or use of *chemicals*, fuel, drugs, food, and many other products. Indeed, chemical engineers apply mathematics, chemistry and other natural sciences to develop practical ways of using energy and materials for the benefit of society. I love to say we make relevant and impactful chemistry and processes happen. But what about the soft skills? This is where Emotional Intelligence becomes a must have. In this presentation, we will highlight how emotional intelligence or the ability to identify, understand and manage emotions is critical in STEM fields and especially during times of crisis. We will explain how self-awareness and self-regulation, intrinsic motivation, empathy and social skills can help you go further in your career and are must haves for managers and leaders of today.

618409 Enhancing Leadership and Management of Individuals and Teams through Emotional Intelligence

Abstract:

Academic performance and job opportunities in engineering and science positions typically place a premium on intelligence (IQ) and analytical ability. Although these are critical competencies for successful employees and leaders/managers, they are often not the determining factor for success in technical positions. Leaders and managers must be able to understand, interpret, and use their and others' emotions to attain goals and function effectively personally and on teams; this requires emotional intelligence (EI or EQ). Traditional engineering or science programs do not address or even mention EI, although the lack of this skill set is evident during specific course requirements and in various career assignments: whenever teams must function efficiently and effectively. The absence of this trait generally leads to frustration and anxiety for leaders and followers and inhibits many competent and motivated engineers and scientists from pursuing or remaining in leadership/management positions. This talk will describe characteristics of technically-trained individuals who lack EI and the effect that this deficiency has on the individual, on team effectiveness and on the attitudes of subordinates, peers, and superiors. Approaches to improve or develop EI and thereby enhance leadership/management prowess will be discussed.

619090 Best Practices Driving Excellence in Plant Performance

Abstract:

Chemical and Process Plants are becoming increasingly more complex as smart equipment is becoming the new industry norm. Operators, Mechanics and Technical Personnel are constantly inundated with overwhelming streams of data. The growth of plant data has to some extent desensitized our staff to the intended warning flags through this data overload, unfortunately allowing critical information to pass through the system unnoticed. So, despite the sophistication of equipment and monitoring systems, plant failures are still too commonplace. As such, site performance continues to fall short of expectations as evidenced by unacceptable unplanned plant downtime, loss of containment related incidents, operator work-arounds due to equipment unavailability, and high loss profit opportunities. As expected, a retrospective analysis often in the form of event investigations most often reveal that existing plant data and "weak signals" were evident well in advance of the impending failures. Typically, readily available information and precursor events were either not detected, ignored or acted upon. Operating and technical personnel were unable to clearly identify, prioritize, communicate, and mitigate the growing plant risk to prevent the undesired "loss of control" and asset damage.

HOPE Consulting personnel with our broad experience from military, national research labs, commercial nuclear power generation, oil & gas, and mining industries has developed and implemented cross-industry best practices to address these performance shortfalls. The goal of our practices is to support uninterrupted runs between turnarounds, and greatly reduce the likelihood of significant equipment and human failures. These methods assist the operating staff to sort through the live plant data and plant physical anomalies to clearly identify, prioritize and mitigate impending equipment/plant failures. In addition, our process allows your organization to calculate the return on investments (ROI) resulting from our proven methods.

Through the integration of critical plant data, the operating teams can review & analyze performance, identify health threats & adverse trends, prioritize them appropriately and implement timely corrective actions. Team members are trained and become proficient in recognizing these "weak signals" and acting upon them before the equipment conditions become self-revealing e.g. equipment failure (enhancing equipment reliability). Personnel are trained in concept of "building the capacity to fail safe" increasing the likelihood that small failures do not cascade into major losses or incidents. Additionally operations, engineering, and maintenance personnel are mentored in the performance of Asset Integrity Inspections, a lost art as staff become increasingly dependent on process data and plant automation while becoming more disconnected from their plant equipment they maintain and operate.

Through this proven process, work teams have demonstrated their ability to proactively identify degrading conditions and underlying causes and enhanced prevention, rather than reacting to plant events. Additionally, business areas can now more effectively manage the supporting reliability programs and reduce the total number of exceptions and deviations which have contributed to unreliability in the past. Finally, systems and management processes are put in place to facilitate the necessary reinforcement, accountability, and process sustainability.

AUTHOR INDEX